For Amelia

✏

Author's Note

Organizing things into categories can range from the most complex of sciences to the totally random, and in this book I have covered pretty much everything in between. On the pages where there is general agreement about how particular things are organized, I have tried as best I can to follow the consensus, but occasionally I have just had to follow my own personal likes and quirky ways of organizing things. Above all, I have tried to make this book beautiful and entertaining. So please forgive any mistakes you might find, and perhaps you could even reorganize some of it in your own way.

First US edition 2020
First published by Walker Books Ltd. (UK) 2020

Library of Congress Catalog Card
Number pending
ISBN 978-1-5362-1121-4

20 21 22 23 24 25 LGO 10 9 8 7 6 5 4 3 2 1

Printed in Vicenza, Italy

This book was typeset in Gill Sans and Filosofia.
The illustrations were done in mixed media.

Candlewick Studio
an imprint of
Candlewick Press
99 Dover Street
Somerville, Massachusetts 02144

www.candlewickstudio.com

ONE
OF A
KIND

A Story About Sorting and Classifying

Neil Packer

CANDLEWICK STUDIO
an imprint of Candlewick Press

This is Arvo.

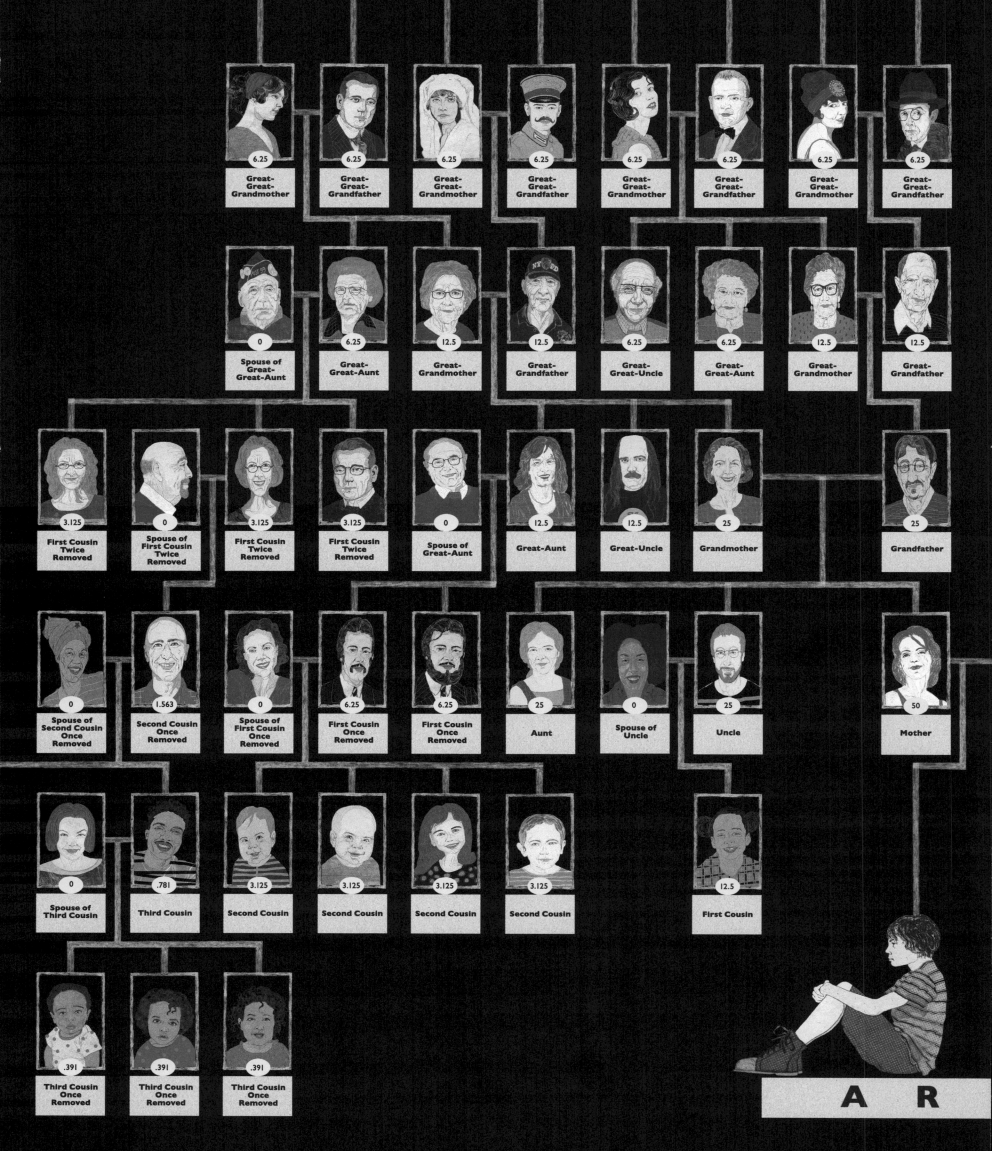

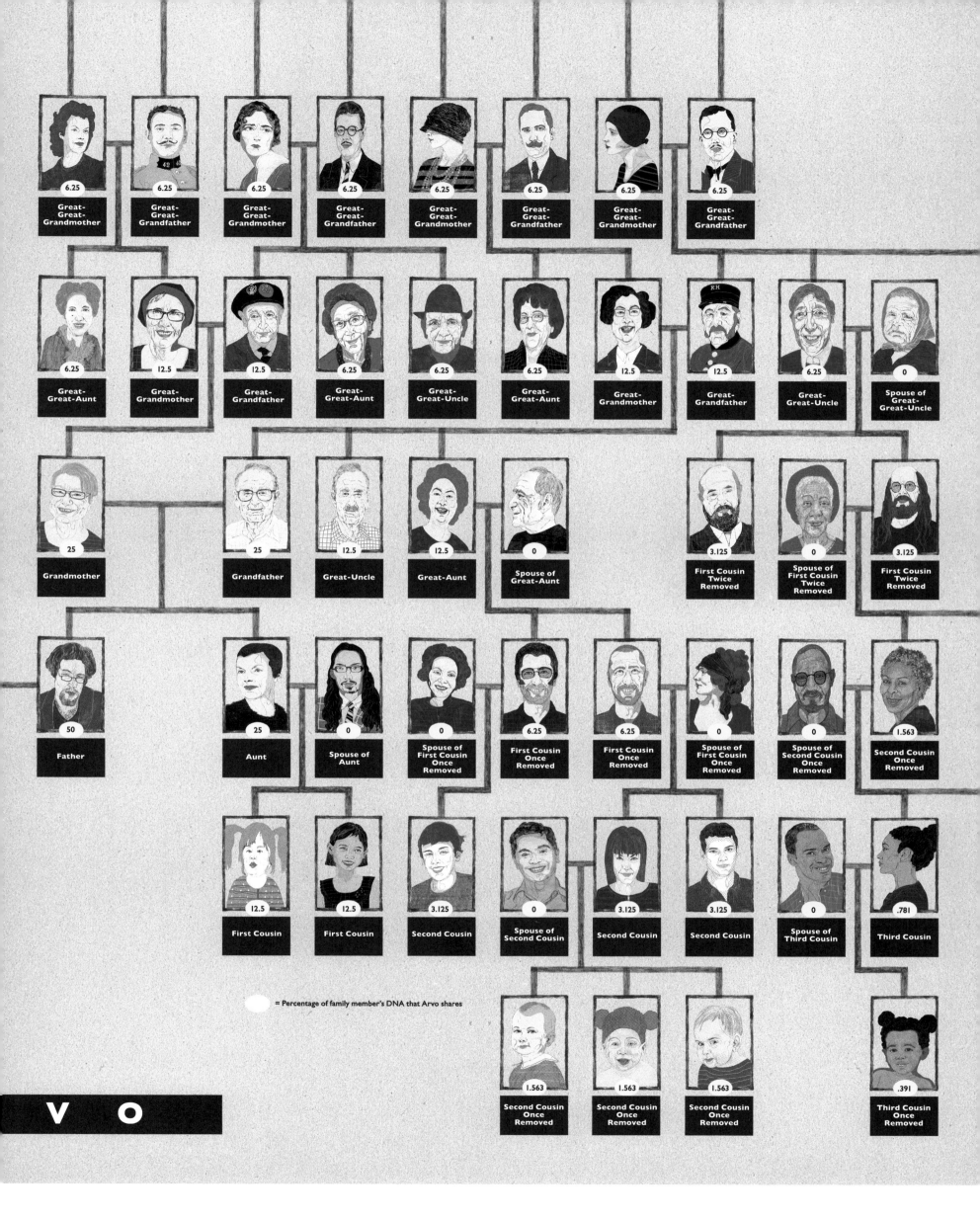

= Percentage of family member's DNA that Arvo shares

And this is Arvo's family.

This is Arvo's cat, Malcolm.

FELIDAE
The family of cats

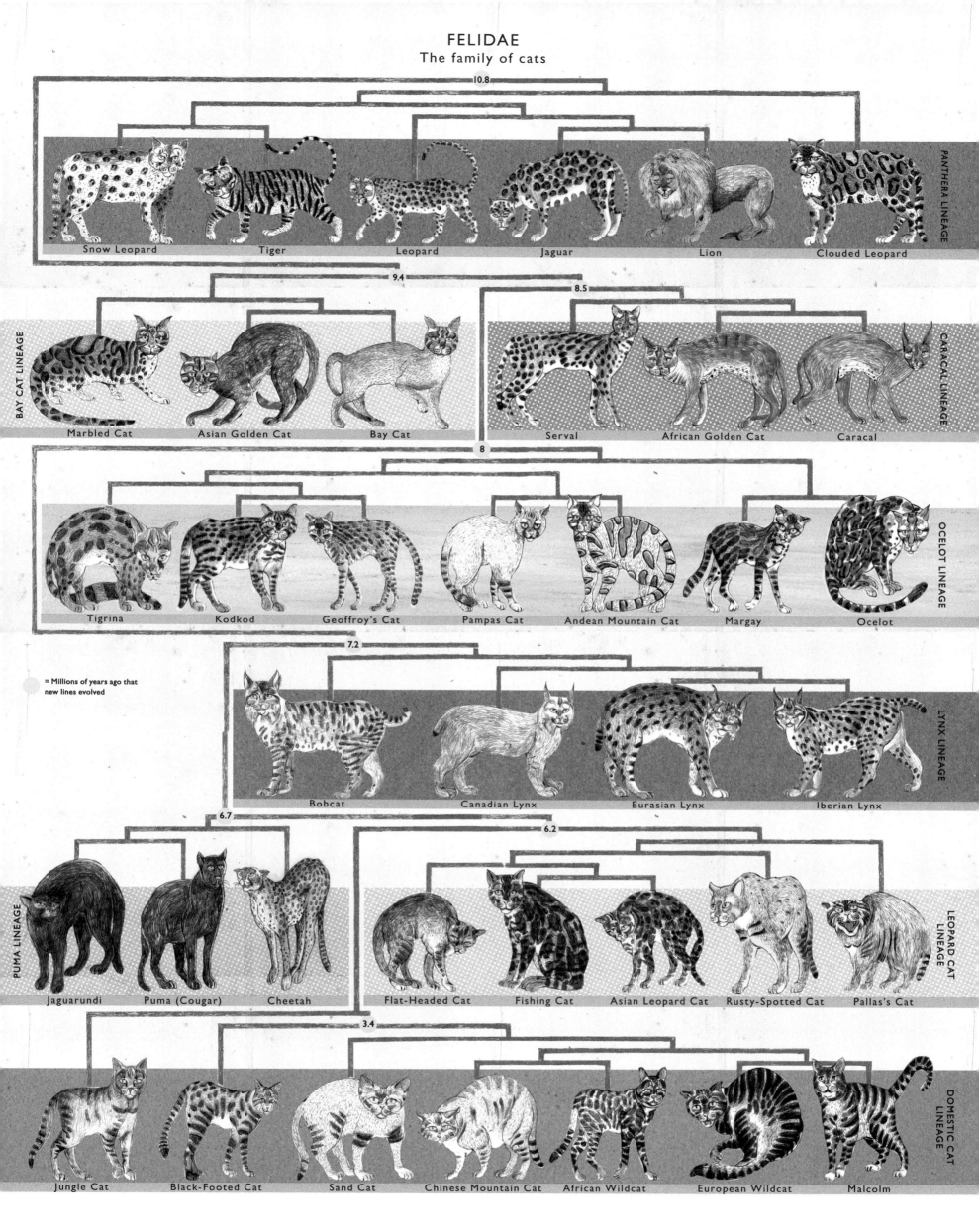

10.8

PANTHERA LINEAGE

Snow Leopard Tiger Leopard Jaguar Lion Clouded Leopard

9.4

8.5

BAY CAT LINEAGE

Marbled Cat Asian Golden Cat Bay Cat

CARACAL LINEAGE

Serval African Golden Cat Caracal

8

OCELOT LINEAGE

Tigrina Kodkod Geoffroy's Cat Pampas Cat Andean Mountain Cat Margay Ocelot

7.2

= Millions of years ago that
new lines evolved

LYNX LINEAGE

Bobcat Canadian Lynx Eurasian Lynx Iberian Lynx

6.7

6.2

PUMA LINEAGE

Jaguarundi Puma (Cougar) Cheetah

LEOPARD CAT LINEAGE

Flat-Headed Cat Fishing Cat Asian Leopard Cat Rusty-Spotted Cat Pallas's Cat

3.4

DOMESTIC CAT LINEAGE

Jungle Cat Black-Footed Cat Sand Cat Chinese Mountain Cat African Wildcat European Wildcat Malcolm

And this is Malcolm's family.

But Arvo and his cat belong to an even bigger group:
the animal kingdom.

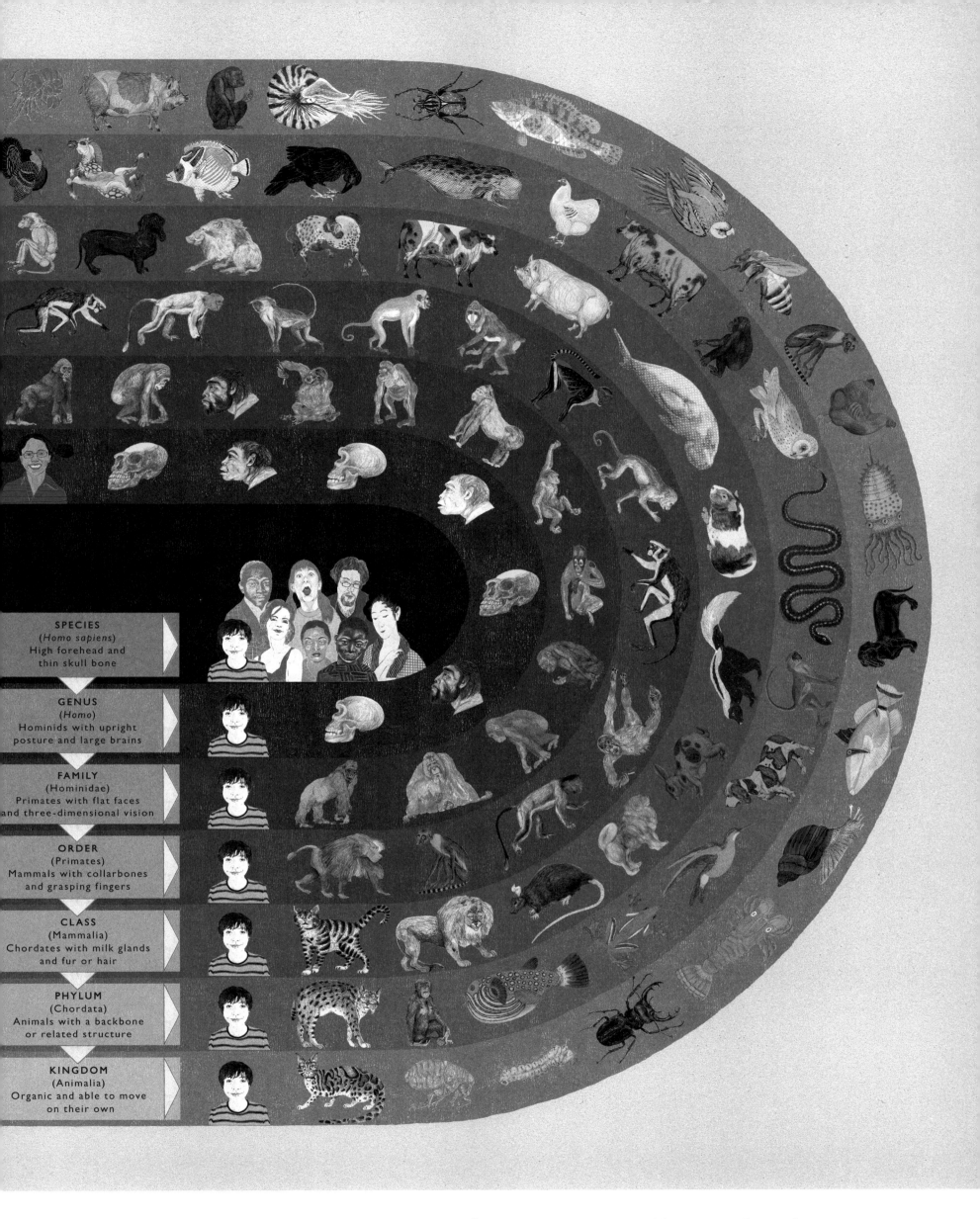

SPECIES
(Homo sapiens)
High forehead and
thin skull bone

GENUS
(Homo)
Hominids with upright
posture and large brains

FAMILY
(Hominidae)
Primates with flat faces
and three-dimensional vision

ORDER
(Primates)
Mammals with collarbones
and grasping fingers

CLASS
(Mammalia)
Chordates with milk glands
and fur or hair

PHYLUM
(Chordata)
Animals with a backbone
or related structure

KINGDOM
(Animalia)
Organic and able to move
on their own

If you are a human, you are one of over seven billion other humans—
but we are only a tiny part of the animal kingdom.

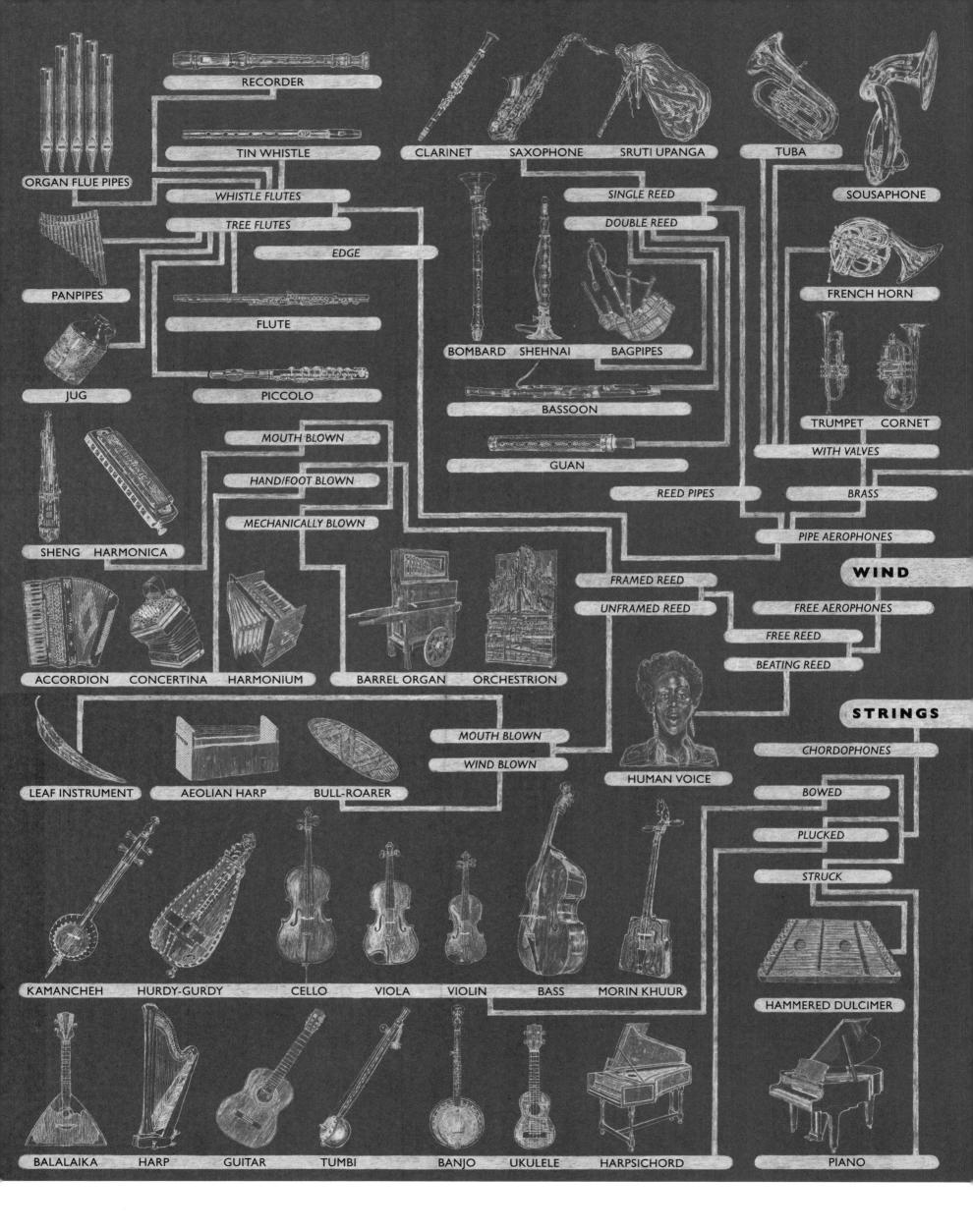

RECORDER

ORGAN FLUE PIPES

TIN WHISTLE

CLARINET   SAXOPHONE   SRUTI UPANGA

TUBA

WHISTLE FLUTES

SINGLE REED

SOUSAPHONE

TREE FLUTES

DOUBLE REED

PANPIPES

EDGE

FRENCH HORN

FLUTE

JUG

BOMBARD   SHEHNAI   BAGPIPES

PICCOLO

BASSOON

TRUMPET   CORNET

MOUTH BLOWN

WITH VALVES

GUAN

HAND/FOOT BLOWN

REED PIPES

BRASS

MECHANICALLY BLOWN

PIPE AEROPHONES

SHENG   HARMONICA

WIND

FRAMED REED

UNFRAMED REED

FREE AEROPHONES

FREE REED

ACCORDION   CONCERTINA   HARMONIUM

BARREL ORGAN   ORCHESTRION

BEATING REED

STRINGS

MOUTH BLOWN

CHORDOPHONES

WIND BLOWN

LEAF INSTRUMENT   AEOLIAN HARP   BULL-ROARER

HUMAN VOICE

BOWED

PLUCKED

STRUCK

KAMANCHEH   HURDY-GURDY   CELLO   VIOLA   VIOLIN   BASS   MORIN KHUUR

HAMMERED DULCIMER

BALALAIKA   HARP   GUITAR   TUMBI   BANJO   UKULELE   HARPSICHORD

PIANO

You can sort everything into groups!

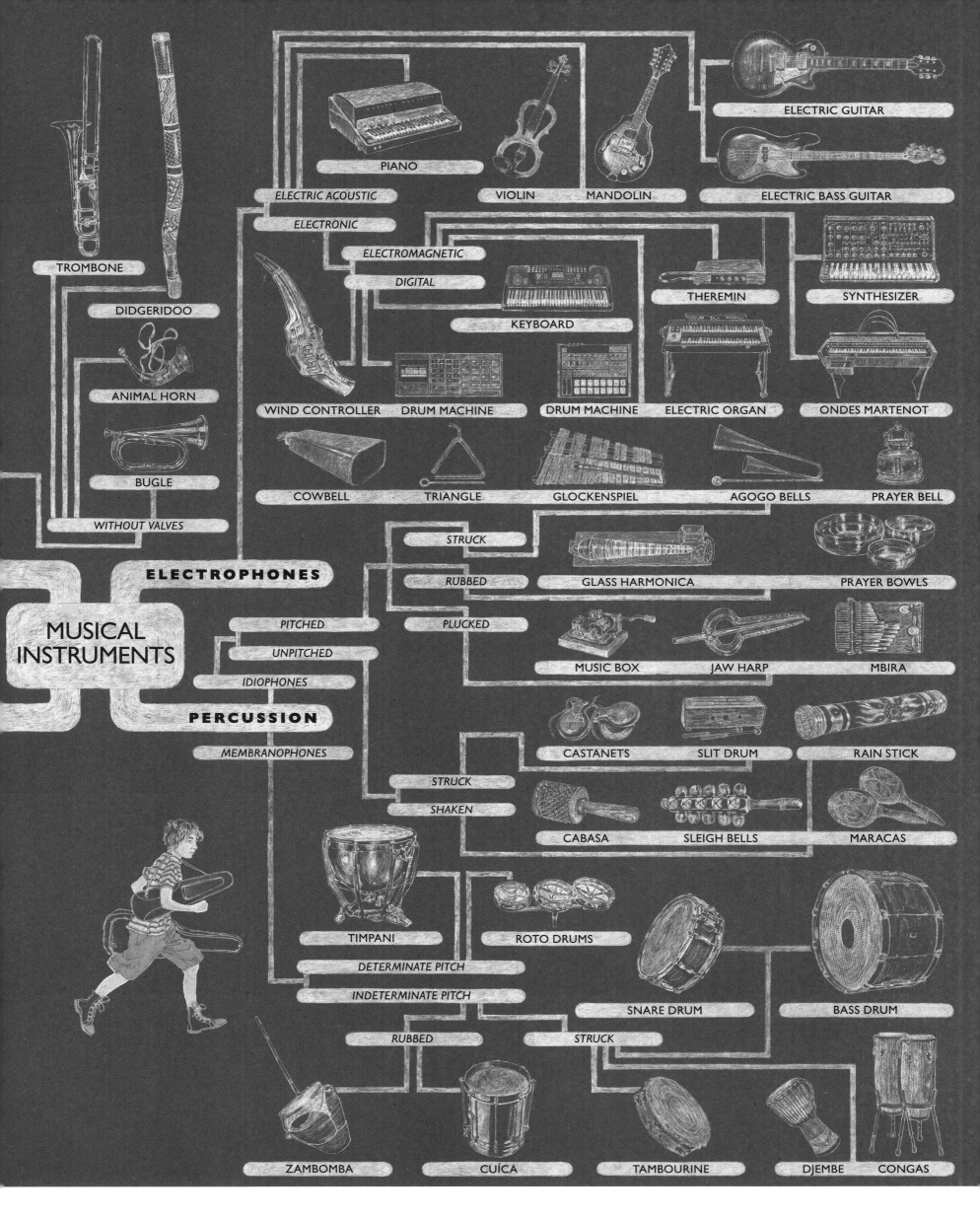

Arvo is learning to play the violin and the guitar. Can you see where they fit into this grouping of musical instruments?

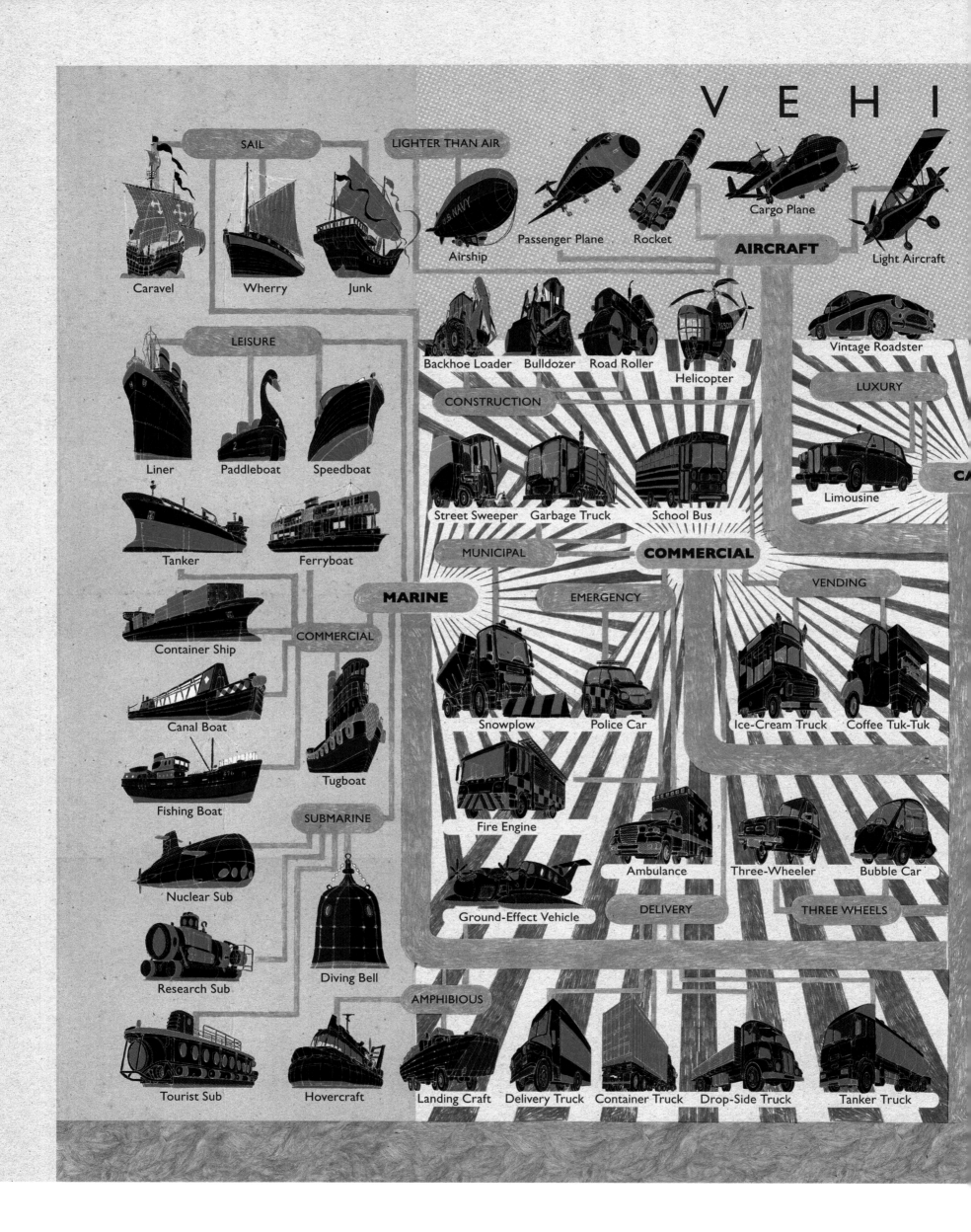

After school, Arvo's dad takes him to his violin lesson in the car.

C L E S

AIRCRAFT

Bomber

VI Doodlebug

COMBAT

STEALTH

Stealth Bomber

TRANSPORT

Tilt Rotor

Long-Range Missile

Jet Fighter

Drone

Helicopter Gunship

Airborne Assault

Air Crane

City Runabout

NASCAR Racer

Tank

Armored Car

Missile Launcher

ELECTRIC

Go-Kart

MILITARY

Mine Clearer

RACING

Mounted Gun Pickup Truck

Four-Wheel Drive Utility

FAMILY

Hatchback

Station Wagon

Local

High Speed

Tank Wagon

Sports Utility Vehicle

PASSENGER

Grain Wagon

Monorail

Subway

FREIGHT

Container Wagon

Tram

Cable Car

Atmospheric Railway

RAIL

TWO WHEELS

Rack-and-Pinion Railway

Maglev

Ballast Regulator

Moped

Child's Scooter

Bicycle

Motorbike

Gyrocar

Arvo's Family Car

The car belongs in a bigger group, too: it's a vehicle,
something we use to travel.

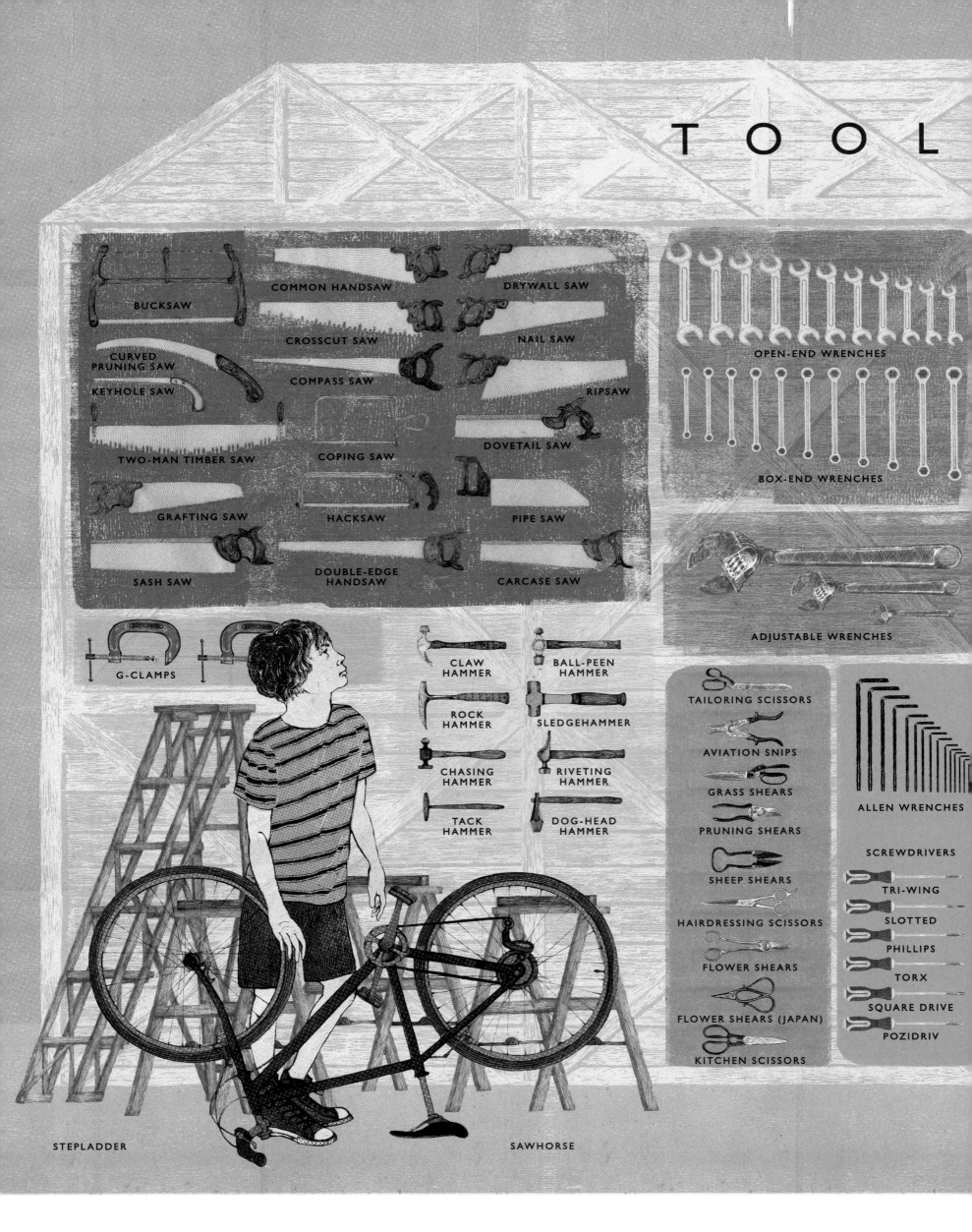

BUCKSAW

COMMON HANDSAW

DRYWALL SAW

CROSSCUT SAW

NAIL SAW

CURVED
PRUNING SAW

COMPASS SAW

KEYHOLE SAW

RIPSAW

TWO-MAN TIMBER SAW

COPING SAW

DOVETAIL SAW

GRAFTING SAW

HACKSAW

PIPE SAW

SASH SAW

DOUBLE-EDGE
HANDSAW

CARCASE SAW

OPEN-END WRENCHES

BOX-END WRENCHES

ADJUSTABLE WRENCHES

G-CLAMPS

CLAW
HAMMER

BALL-PEEN
HAMMER

TAILORING SCISSORS

ROCK
HAMMER

SLEDGEHAMMER

AVIATION SNIPS

ALLEN WRENCHES

CHASING
HAMMER

RIVETING
HAMMER

GRASS SHEARS

SCREWDRIVERS

PRUNING SHEARS

TRI-WING

TACK
HAMMER

DOG-HEAD
HAMMER

SHEEP SHEARS

SLOTTED

HAIRDRESSING SCISSORS

PHILLIPS

FLOWER SHEARS

TORX

SQUARE DRIVE

FLOWER SHEARS (JAPAN)

POZIDRIV

STEPLADDER

SAWHORSE

KITCHEN SCISSORS

Here's Arvo's bicycle—another kind of vehicle.

SHED

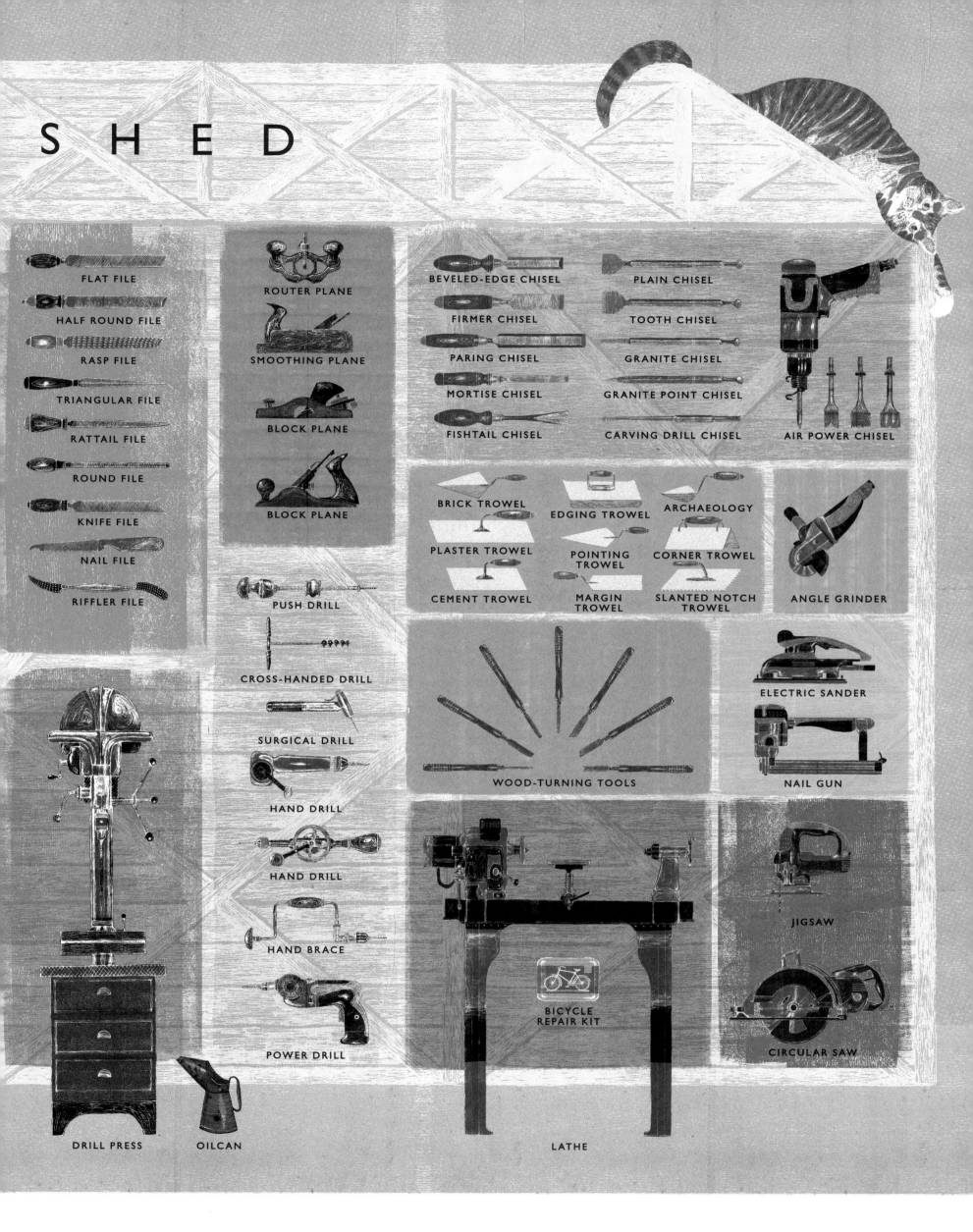

FLAT FILE

HALF ROUND FILE

RASP FILE

TRIANGULAR FILE

RATTAIL FILE

ROUND FILE

KNIFE FILE

NAIL FILE

RIFFLER FILE

ROUTER PLANE

SMOOTHING PLANE

BLOCK PLANE

BLOCK PLANE

PUSH DRILL

CROSS-HANDED DRILL

SURGICAL DRILL

HAND DRILL

HAND DRILL

HAND BRACE

POWER DRILL

DRILL PRESS

OILCAN

BEVELED-EDGE CHISEL

FIRMER CHISEL

PARING CHISEL

MORTISE CHISEL

FISHTAIL CHISEL

PLAIN CHISEL

TOOTH CHISEL

GRANITE CHISEL

GRANITE POINT CHISEL

CARVING DRILL CHISEL

AIR POWER CHISEL

BRICK TROWEL

EDGING TROWEL

ARCHAEOLOGY

PLASTER TROWEL

POINTING TROWEL

CORNER TROWEL

CEMENT TROWEL

MARGIN TROWEL

SLANTED NOTCH TROWEL

ANGLE GRINDER

WOOD-TURNING TOOLS

ELECTRIC SANDER

NAIL GUN

BICYCLE REPAIR KIT

LATHE

JIGSAW

CIRCULAR SAW

It's got a flat tire. Which tools would be best to fix it?

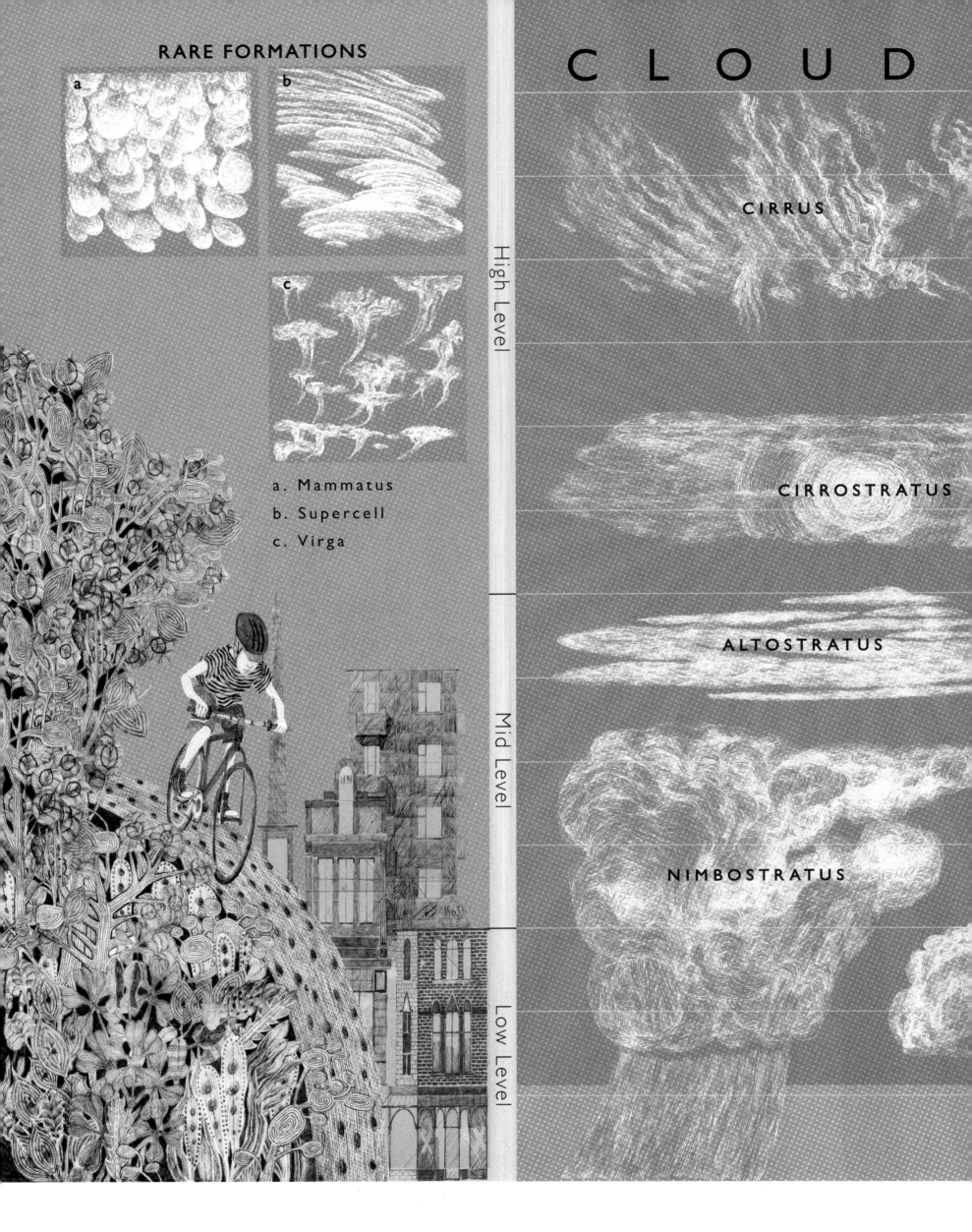

RARE FORMATIONS

a. Mammatus
b. Supercell
c. Virga

CLOUD

CIRRUS

CIRROSTRATUS

ALTOSTRATUS

NIMBOSTRATUS

High Level

Mid Level

Low Level

When he's fixed his bike, Arvo rides into town.

TYPES

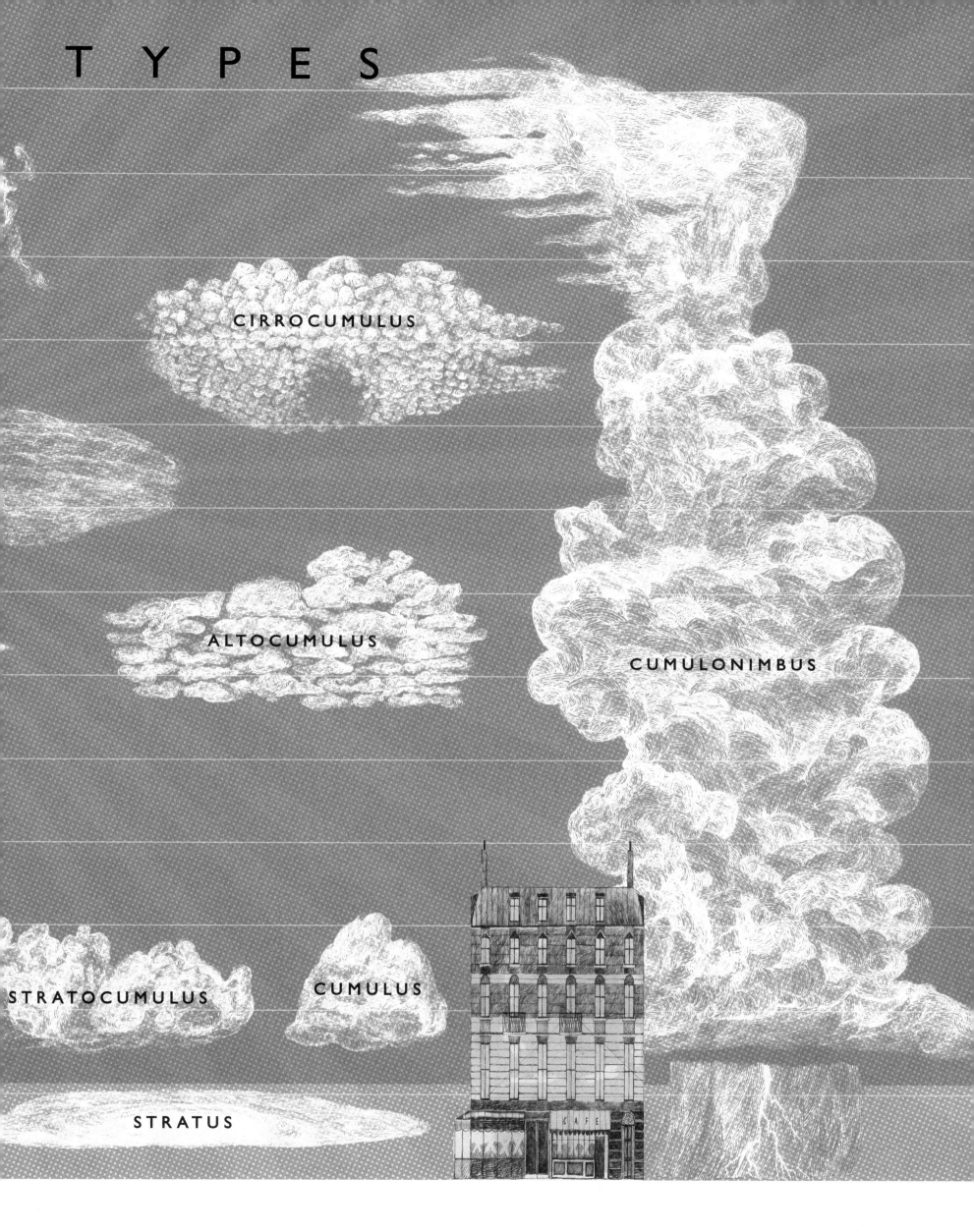

CIRROCUMULUS

CUMULONIMBUS

ALTOCUMULUS

STRATOCUMULUS

CUMULUS

STRATUS

Better hurry, Arvo—bad weather's coming.

The buildings in Arvo's town can be sorted into different groups:

according to what they're used for . . .

3,000 YEARS AGO   **400** YEARS AGO

100 YEARS AGO

300
YEARS AGO

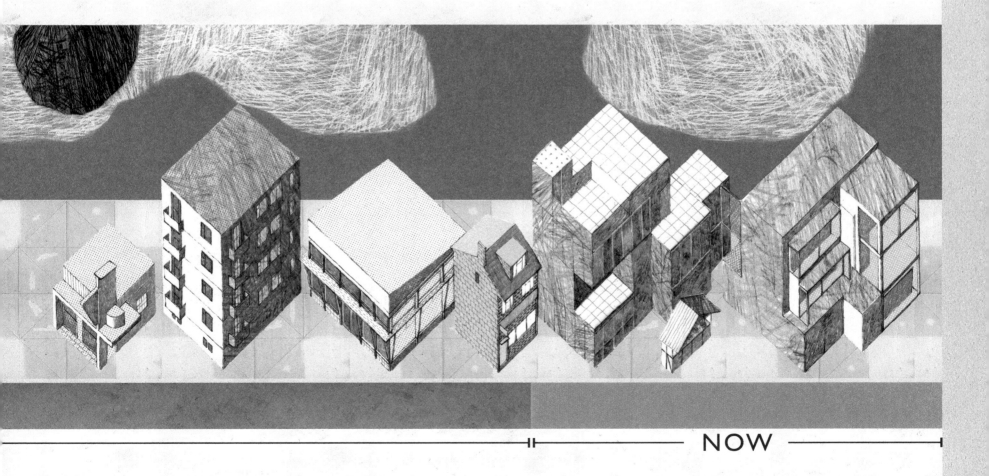

NOW

how old they are . . .

M A T E R I A L S

MOSTLY STONE

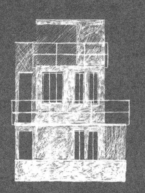

MOSTLY CONCRETE

MOSTLY METAL

MOSTLY GLASS

MOSTLY BRICK

MOSTLY WOOD

or what they're made of.

Arvo heads for the farmers' market.

His dad asked him to buy some apples—there are plenty
of different kinds to choose from!

THE ARTS

RECREATION

LITERATURE & FICTION

Performing Arts

History of Art

Sports / Games

General

Classics

Architecture

Painters

Sport & Leisure Industries

Adventure

Horror

Drawing / Decorative Arts

Graphic Design

Quotations

Romance

Photography

APPLIED SCIENCES

Humor

Science Fiction

Music

Technology

Poetry

Literary Criticism

Medicine

Engineering

Europe

Africa

Manufacturing

Nature

Ancient World

The Americas

Mathematics

Astronomy

Asia

Physics

Chemistry

Biography

Biology

Dinosaurs

History Journals

British Isles

Animals

Travel

History

Plants

General Resources

Graphic Novels

NATURAL SCIENCES & MATHEMATICS

GEOGRAPHY & HISTORY

Next Arvo pops into the library. He's looking for a book about art,

LIBRARY

GENERAL KNOWLEDGE

PHILOSOPHY & PSYCHOLOGY

RELIGION

Computing / The Internet

Modern Western Philosophy

Christianity

Islam

Knowledge / Research

Ancient Philosophy

Judaism

Hinduism

Systems

Psychology

Sociology

Buddhism

Bibliography

Economics

SOCIAL SCIENCES

Logic

News Media / Publishing

Ethics

Political Science

Law

Reference Books

Education

General

Study Guides

Religion / Spirituality

Customs / Folklore

Geography

Society

Work and Money

Personal Issues

Science / Math / Technology

Literature

EARLY YEARS

Animals

Poetry

Books for Babies

Art / Music / Drama

Humor

Early Learning

Sports

Concept Books

CHILDREN & YOUNG ADULTS

Picture Books

and it's easy to find because the books are all grouped by subject.

INSULAR ART
Celtic & Anglo-Saxon, 600 CE

PERSIAN MINIATURE
13th Century

RENAISSANCE
Europe, 15th–16th Centuries

UKIYO-E
Japanese Woodblock Prints, 1600

IMPRESSIONISM
Paris, 1872

EXPRESSIONISM
Germany, Late 19th Century

CUBISM
France, 1907

CONSTRUCTIVISM
Russia, 1913

NUBIAN WALL DECORATION
Egypt / Sudan, 1920s–1964

SOCIALIST REALISM
Russia, Late 1920s

ABSTRACT EXPRESSIONISM
USA, Late 1940s

MANGA
Japan, Mid-20th Century

The paintings in the book are amazing—all so different.

DUTCH GOLDEN AGE
Netherlands, 17th Century

FOLK ART (LUBOK)
Russia, Mid-17th Century

ROCOCO
France, 18th Century

ROMANTICISM
Europe, 1780s–1850

FOUND OBJECT
France & USA, 1917

DE STIJL
Holland, 1917

SURREALISM
Europe, Early 1920s

ART DECO
Europe, 1920s

POP ART
USA / UK, Mid 1950s

MINIMALISM
USA, 1960s

NEO-EXPRESSIONISM
Late 1970s

POSTMODERNIST
Late 1970s

But even they can be sorted into groups.

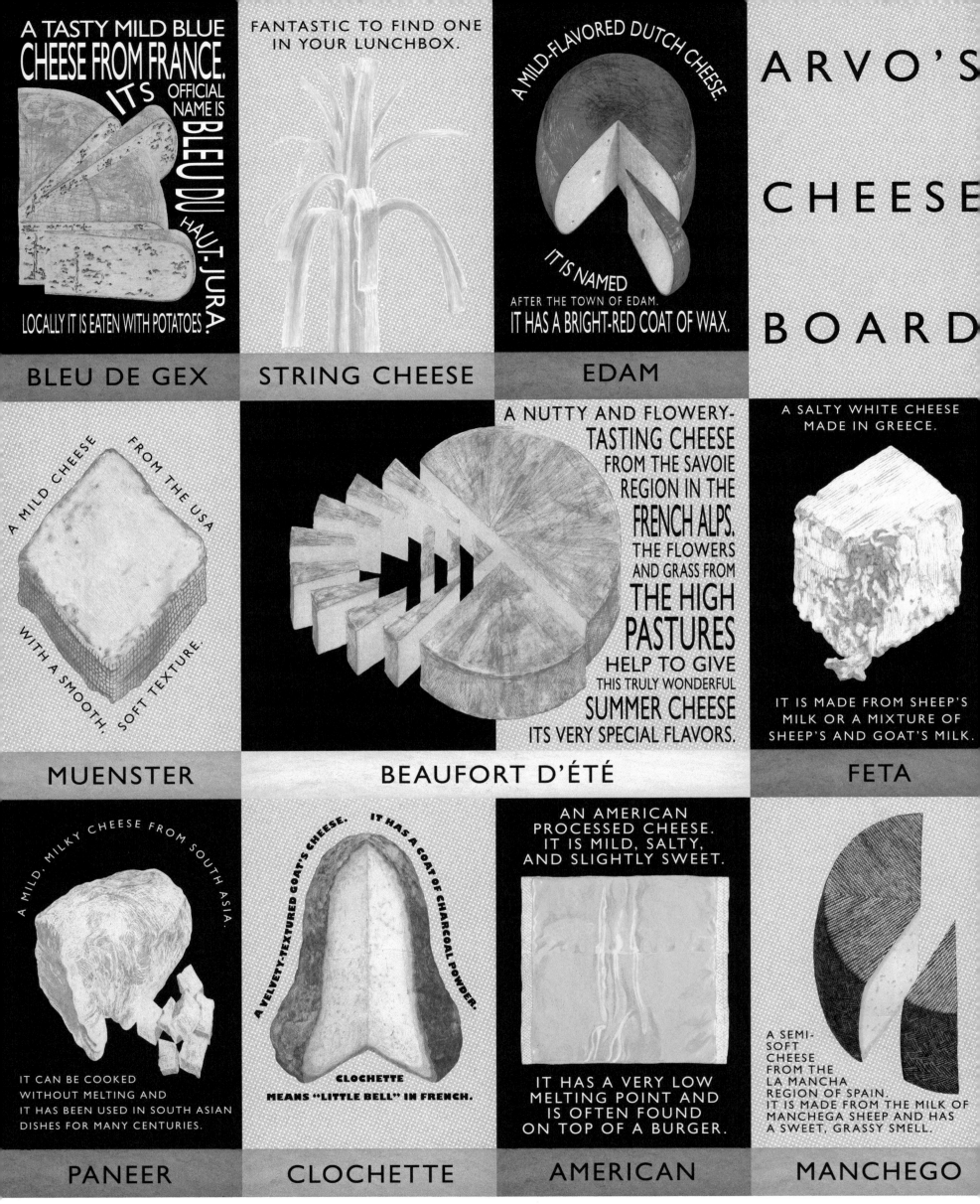

BLEU DE GEX — A TASTY MILD BLUE CHEESE FROM FRANCE. ITS OFFICIAL NAME IS BLEU DU HAUT-JURA. LOCALLY IT IS EATEN WITH POTATOES.

STRING CHEESE — FANTASTIC TO FIND ONE IN YOUR LUNCHBOX.

EDAM — A MILD-FLAVORED DUTCH CHEESE. IT IS NAMED AFTER THE TOWN OF EDAM. IT HAS A BRIGHT-RED COAT OF WAX.

ARVO'S CHEESE BOARD

MUENSTER — A MILD CHEESE FROM THE USA WITH A SMOOTH, SOFT TEXTURE.

BEAUFORT D'ÉTÉ — A NUTTY AND FLOWERY-TASTING CHEESE FROM THE SAVOIE REGION IN THE FRENCH ALPS. THE FLOWERS AND GRASS FROM THE HIGH PASTURES HELP TO GIVE THIS TRULY WONDERFUL SUMMER CHEESE ITS VERY SPECIAL FLAVORS.

FETA — A SALTY WHITE CHEESE MADE IN GREECE. IT IS MADE FROM SHEEP'S MILK OR A MIXTURE OF SHEEP'S AND GOAT'S MILK.

PANEER — A MILD, MILKY CHEESE FROM SOUTH ASIA. IT CAN BE COOKED WITHOUT MELTING AND IT HAS BEEN USED IN SOUTH ASIAN DISHES FOR MANY CENTURIES.

CLOCHETTE — A VELVETY-TEXTURED GOAT'S CHEESE. IT HAS A COAT OF CHARCOAL POWDER. CLOCHETTE MEANS "LITTLE BELL" IN FRENCH.

AMERICAN — AN AMERICAN PROCESSED CHEESE. IT IS MILD, SALTY, AND SLIGHTLY SWEET. IT HAS A VERY LOW MELTING POINT AND IS OFTEN FOUND ON TOP OF A BURGER.

MANCHEGO — A SEMI-SOFT CHEESE FROM THE LA MANCHA REGION OF SPAIN. IT IS MADE FROM THE MILK OF MANCHEGA SHEEP AND HAS A SWEET, GRASSY SMELL.

On his way out, Arvo stops to look

A HARD, STRONG-FLAVORED CHEESE FROM ITALY. IT IS OFTEN GRATED OR SHAVED OVER PASTA DISHES.

A TASTY AND SEMIHARD ENGLISH CHEESE MADE FROM COW'S MILK.

IT IS NAMED AFTER THE VILLAGE OF CHEDDAR WHERE IT WAS FIRST MADE. IT IS ONE OF THE WORLD'S MOST POPULAR CHEESES.

A SOFT FRESH CHEESE FROM ITALY, TRADITIONALLY MADE FROM BUFFALO MILK.

A MATURE VERSION OF MONTEREY JACK. IT HAS A SWEET, NUTTY FLAVOR.

CHEDDAR

MOZZARELLA

DRY JACK

ANOTHER GREAT LUNCH-BOX TREAT. SPREADABLE CHEESE TRIANGLES WERE INVENTED NEARLY 100 YEARS AGO. THERE ARE MANY DIFFERENT TYPES AND THEY CAN BE FOUND ALL OVER THE WORLD.

A WHITE SALTY CHEESE POPULAR IN GREECE, TURKEY, CYPRUS, AND THE LEVANT.

IT CAN BE GRILLED, FRIED, OR BAKED WITHOUT MELTING.

PARMESAN

PROCESSED CHEESE TRIANGLES

HALLOUMI

A MILKY, SWEET-TASTING FRENCH CHEESE. IT IS MADE IN NORMANDY FROM THE RICH MILK OF SPECKLED BROWN OR BLACK NORMANDE COWS.

A FRUITY-FLAVORED HARD CHEESE FROM SWITZERLAND WITH A SWEET AND NUTTY SMELL.

EMMENTAL

IT IS FAMOUS FOR ITS HOLES.

A MILD-TASTING SOFT CHEESE MADE FROM

COW'S MILK AND CREAM. IT IS SHOWN HERE SPREAD ON A CRACKER.

CAMEMBERT

EMMENTAL

CREAM CHEESE

at a book about his favorite food—cheese!

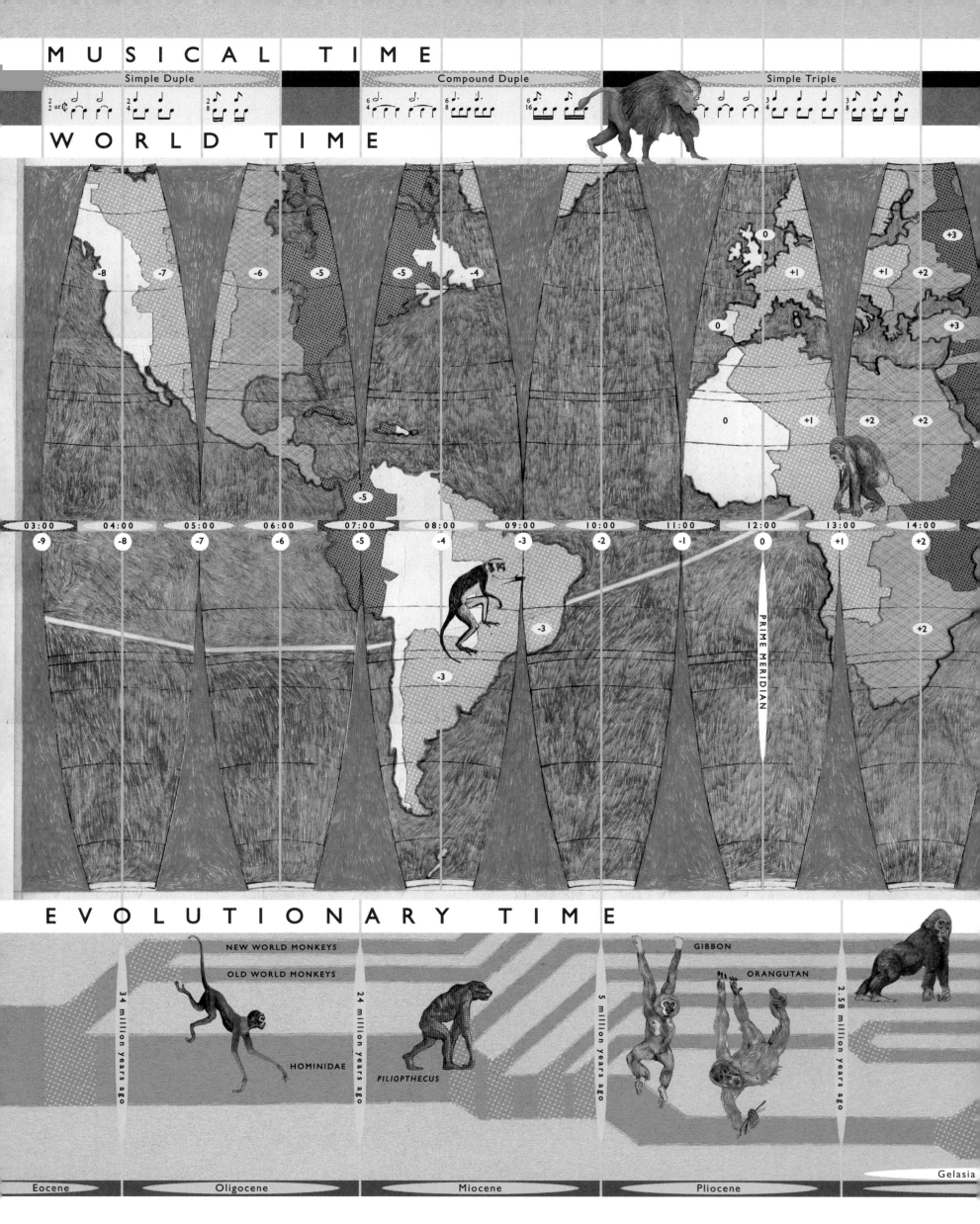

Arvo spends a long time in the library—
much longer than he'd planned.

How long is a long time?

Arvo's worried that he might be late getting home.
But then he spots a familiar face smiling at him. It's his dad.

Arvo asks him how it can possibly be—that out of all the billions of people in the world, they can always pick each other out of a crowd.

"That's easy," says Dad. "There are lots of people in the world, but there's only one ME . . .

"And only one

YOU!"

Sorting things into groups is how we make sense of the world. Imagine trying to find a book in a library if someone hadn't gone to the trouble of putting the books into an order, or trying to find something in a supermarket if everything was all jumbled up!

But it's more than just that—in science, organizing things has led to major breakthroughs. The periodic table is a good example: it was Russian scientist Dmitri Mendeleev's way of organizing the pure materials (or "elements") of the world into an order that made sense. By studying the pattern of similarities between elements, Mendeleev predicted the existence of materials that were yet to be discovered.

I have always been fascinated by the idea of sorting things into categories. When I was a child, I especially liked to draw different vehicles and then divide them into neat groups; I would ask myself questions like: Did they travel by land, water, or air? Did they carry people or cargo? And so on.

But I quickly learned that my vehicles could be arranged in all sorts of different ways; they could fit in more than one category, and belong to several groups at once. It wasn't an exact science.

That's something I wanted to explore in this book: that the categories we invent are never perfect. And although it is useful and interesting to order things, it is often equally important *not* to. Creativity isn't always orderly—and disorder can sometimes be beautiful.

Neil Packer

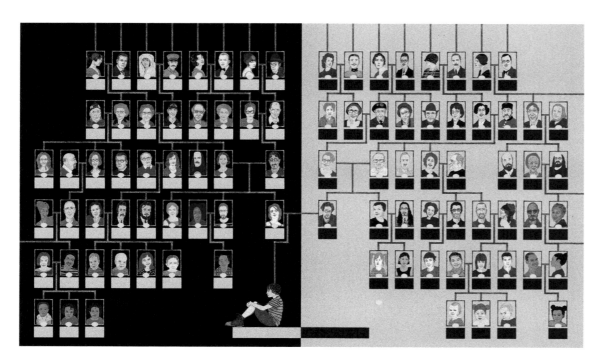

1 – Arvo's Family Tree

"Family tree" is the name for a diagram showing the members of a family and how they are related to one another.

There are different ways to represent this, but family trees often show a single person at the bottom. The relationships between that individual and their extended family then branch out above them, going back over several generations.

Trees have been used to symbolize family history for many centuries. The longest family tree in the world is that of the Chinese philosopher Confucius, who died in 479 BCE. It includes eighty generations and more than two million family members.

In this book, Arvo can only trace his family back a few generations—perhaps that's just as well, or I might still be working on it now! ∎

2 – Malcolm's Family Tree

Malcolm's family tree is quite different from Arvo's; it doesn't show the members of Malcolm's immediate family, like his brothers, sisters, and parents.

Instead this picture is about the evolution of the cat family, also called *Felidae*. Here, "family" means a grouping within the larger animal kingdom (more on that later). And *evolution* is the word we use to describe how, over long periods of time, an animal family changes as species develop adaptations to their surroundings.

This happens when a particular genetic characteristic, such as, say, superior night

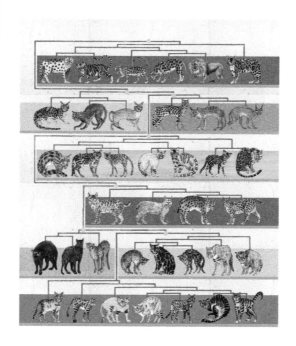

vision, results in a lineage of animals being successful in their environment.

Over millions of years, new species of cats have evolved from ancient species like the lion and the tiger, eventually resulting in the domestic cat—that is, Malcolm! ∎

3 – The Animal Kingdom

Every plant and animal on Earth can be categorized using the Linnaean system. This method was developed by Swedish scientist Carl Linnaeus, who lived during the eighteenth century, a time when science was making huge advances.

Linnaeus created a way to name and order all living things by placing them in groups of ever more distant relatives. Those groups are called "taxonomic ranks"; the word *taxonomic* comes from two Greek words: *taxis*, meaning "arrangement," and *nomia*, meaning "method."

With human beings as an example, this illustration shows Linnaeus's different rankings.

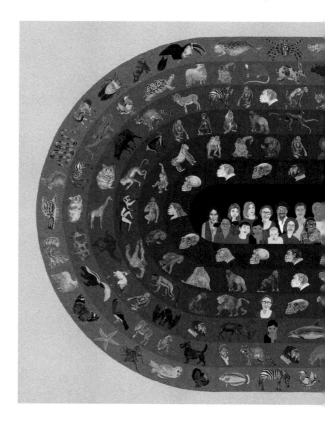

In order of smallest to largest, these are: species, genus, family, order, class, phylum, and kingdom.

All human beings are members of a particular species, *sapiens*, within a particular genus, *Homo*. That's why human beings have the Latin name of *Homo sapiens*.

The next largest rank is family, which in our case is Hominidae, a group that includes the higher apes like chimpanzees and gorillas. Then come order, class, and phylum. Finally, human beings belong to the animal kingdom rather than the plant kingdom or the fungi kingdom. ■

4 — Musical Instruments

Musical instruments are commonly grouped into families. The illustration shows the brass, string, woodwind, percussion, and keyboard families.

Within each family, there are smaller groups of instruments. For instance, the string family is made up of instruments that are plucked, like guitars; bowed, like violins; or even struck, like the hammered dulcimer.

Interesting fact: the saxophone, which is traditionally made of brass, is a woodwind instrument. Meanwhile, the didgeridoo, which is made of wood, is a brass instrument.

Although this may seem odd, it makes sense when you understand how musical instruments are grouped. They are ordered not by what material they are made from but by how they generate their sound.

Like a bugle, the didgeridoo is a hollow tube, which makes it a member of the brass family, and like a clarinet, the saxophone has a reed, making it belong to the woodwind family. ■

5 — Vehicles

This diagram is similar to the one for musical instruments; this time, it shows the many kinds of vehicles there are. We use the word *vehicle* to describe any machine that transports people or cargo—whether by land, sea, air, or even through space.

For this book, I first sorted the vehicles into general categories, such as rail (vehicles that travel by railway), marine (vehicles that travel by water), and so on. I then divided them up as best I could under these headings, whether by number of wheels or by more specific use, such as passenger (vehicles to carry people around) or freight (vehicles to carry objects around).

You might notice that this diagram looks quite like a tree—possibly even more so than the family trees at the beginning of the book. Diagrams like this are sometimes called "trees of life," and I find them to be a good, clear way of showing lots of smaller groups within one big category. ■

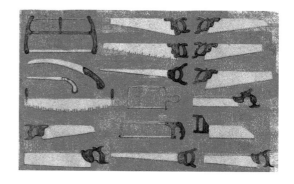

6 — Toolshed

This picture shows the kind of tools you might come across in a mechanic's workshop or a garden shed.

A tool is defined in the dictionary as "an instrument or device, usually held in the hand and used to do a particular job." Tools are an essential part of many people's jobs: engineers, dentists, and artists all use tools in their everyday working life.

As you can see, a single tool like a saw can come in a bewildering number of varieties. They can look very different indeed, depending on their intended purpose. For example, there are saws designed specifically to be used by carpenters, gardeners, and surgeons. ■

7 — Cloud Types

The clouds that Arvo passes under on his bike ride into town are, coincidentally enough, the ten basic types of cloud formations. These are based on the studies of Victorian writer and meteorologist Luke Howard ("meteorologist" being the name for a scientist who studies the weather).

In a book published in 1803, Howard suggested naming the most common types of clouds. Although he wasn't the first person to propose the idea, his system caught on because he used Latin words that can be combined to elegantly describe clouds in all their variations.

Stratus or *strato* means flat, layered, and smooth; *cumulus* or *cumulo* means heaped up and puffy; *cirrus* or *cirro* means high and wispy; *alto* means medium-level; and *nimbus* or *nimbo* means rain-bearing.

The enormous cumulonimbus you see means there's probably a storm brewing! ∎

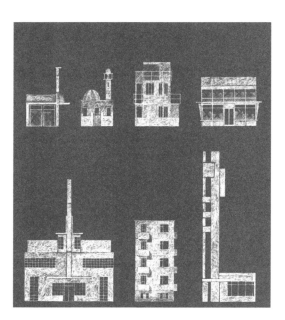

8 – Buildings

There are three different diagrams of buildings in this book. I've included them to show how these groupings we humans come up with aren't always rigid.

In fact, things can be reorganized in all kinds of ways, depending on what we need to understand about a particular topic.

I've organized the buildings in Arvo's town first by use, such as shops, homes, offices; then by how long ago they were built; and finally by

the materials from which they were made.

As you shuffle these categories around, patterns begin to emerge. For example, we can see when concrete became popular as a building material and how styles of architecture changed over a period of time.

These changes can be governed by taste, but also come about as a result of developments in building materials—such as when steel-framed construction made it possible to build taller structures in the late 1800s.

However, Arvo's town, like many others, doesn't have any skyscrapers. ∎

9 – Apples

There are more than 7,000 varieties of apples in the world. And unlike plants that grow in the wild, which slowly adapt to their surroundings over millions of years, the varieties of apples that we buy from the market are carefully bred by farmers over relatively short periods of time.

More often than not, this is to develop a certain aspect of an apple's flavor or appearance. This is not easy because the fruit on a tree grown from the seeds of a single apple tends not to look (or taste) much like the original.

For this reason, farmers have to breed new varieties from two different "parent" apples. Eventually, thousands of trees might be grown from the seeds of the resulting fruit—but only a handful of trees that have the ideal characteristics will be used as the "cultivar" (the trees from which the new variety is grown).

There wasn't room to show all 7,000-and-something varieties of apples in this book, so I have shown only sixty-six.

They are ordered as is usually the case at a market—in a way that is designed to catch the customer's eye!

10 – Books

The library in Arvo's town is quite small, but you can still find a book about almost anything there.

There are different systems to organize books used around the world. But they are all designed to make it easy for people to find the book they want, and they do this by dividing books up into ever more specific categories.

If you want to find a book about slugs for a school nature project, first you need to find the Children's Nonfiction section. Then you could look for the Science section, then the Animals section within that. Then you'd see a section on Mollusks, with books about snails and slugs.

Where might you find this book in a library? (Or perhaps you just have!) ∎

11 – Art

Just by looking at an artwork, you can often tell *when* or *where* it was made. For example, certain styles or materials are popular in particular parts of the world.

But studying art is also a good way of understanding our history, because different styles overlap one another on a time line as well as on a map. Momentous events, changing tastes, and the discovery of new techniques can all affect the way artists work.

Occasionally groups of artists with similar ideas emerge, and the new styles of art they create are referred to as movements.

New movements can evolve slowly out of existing movements, but sometimes they explode into existence very quickly as a result of someone having a radical idea.

For example, in 1913, French artist Marcel Duchamp (who had already spent some time questioning what a work of art could be) had a radical idea. He exhibited everyday items and found objects, or "readymades," as he called them, and announced that they were art.

This illustration shows art movements in rough order of date—and is just a small sample of the many, many different movements that have existed over the years. ∎

12 – Arvo's Cheese Board

There are lots of different ways you could categorize cheese: according to which animal's milk it's made from, where it comes from, or by its age.

When a cheese is first made, it's called a fresh cheese and is soft, like cottage cheese or cream cheese. When it's a bit older, it's semi-soft, like Brie and Camembert.

The next oldest cheeses, like Edam and Emmental, are medium-hard; then come the semi-hard cheeses like Cheddar and Monterey Jack.

Finally, hard cheeses can be several years old, like Parmesan and Dry Jack.

Of course, Arvo's cheeses aren't ordered in any particular sort of way—they're simply his favorites! ∎

13 – Time Map

This map covers three different kinds of time: music time, world time, and evolutionary time.

Musical tempo orders time by dividing it into bars and beats. The most common is 4/4 time, which is four beats to every bar. Most pop, rock, blues, and rap music is written in 4/4 time.

Meanwhile, world time is measured by a twenty-four-hour clock, based on the time it takes for the Earth to turn a full circle on its axis.

The earliest clocks we know of were sundials and water clocks, invented by the ancient Egyptians roughly 3,500 years ago. They, and all the clocks after them, set midday as the point

when the sun is at its highest in the sky, and as this is different in different places, cities only a few miles apart had different times.

When railways were built across Britain in the nineteenth century, every clock in the country had to be set to tell the same time to help coordinate schedules. And in the twentieth century, twenty-four global time zones were put in place, all regulated to a point in London: Greenwich Mean Time, which is now known as Coordinated Universal Time.

Now you can tell the time anywhere in the world by adding or subtracting the hours from Coordinated Universal Time on a time zone map (like the one in this book).

Scientists who study evolution cover billions of years, so measuring time in hours, days, and months is not much use to them.

Instead they divide evolutionary time into large chunks called eras. Each era is divided into periods, and periods are divided into epochs. We humans evolved from our ape ancestors over the course of 2.6 million years, during the Holocene epoch of the Quaternary period of the Cenozoic era.

For our diagram to show the evolution of all life on Earth, from 3.8 billion years ago to the present day, it would have to be very, very long—far too long to fit in this book! ∎

14 – DNA

DNA stands for deoxyribonucleic acid (phew!) and it is found in the cells of every living thing on this planet.

DNA does a job similar to that of computer code: it is the detailed plan for how every part of your body looks and works.

As a human being, you share about 99.9% of your DNA with every other person on the planet—with that 0.1% making a big difference. (You also share about 99% of your DNA with chimpanzees and, on average, 85% with mice!)

DNA is very beautiful: it's shaped like a twisted ladder, called a double helix. Each step or rung on the ladder is made up of four different chemicals, which are called bases.

Each base has its own letter—A, T, G, or C—and the bases are arranged in an order that depends on both of your parents' DNA, but is not identical to theirs.

DNA determines what you look like, but it doesn't define who you are as a person. That's something all of us do for ourselves, through our thoughts and deeds.

How you choose to live your life makes you every bit as special as your own, very unique, DNA. ∎